Real-Life Math

Geometry

by
Walter Sherwood

illustrated by Lois Leonard Stock

User's Guide to *Walch Reproducible Books*

As part of our general effort to provide educational materials which are as practical and economical as possible, we have designated this publication a "reproducible book." The designation means that purchase of the book includes purchase of the right to limited reproduction of all pages on which this symbol appears:

Here is the basic Walch policy: We grant to individual purchasers of this book the right to make sufficient copies of reproducible pages for use by all students of a single teacher. This permission is limited to a single teacher, and does not apply to entire schools or school systems, so institutions purchasing the book should pass the permission on to a single teacher. Copying of the book or its parts for resale is prohibited.

Any questions regarding this policy or requests to purchase further reproduction rights should be addressed to:

Permissions Editor
J. Weston Walch, Publisher
P. O. Box 658
Portland, Maine 04104-0658

1 2 3 4 5 6 7 8 9 10

ISBN 0-8251-4259-8

Printed in the United States of America

Contents

Preface

Picture this: You are a contestant on a game show, and it is down to the final question. Answer it correctly and you win a million dollars. Category is—geometry.

> The host reads the question: "A 10-foot ladder leans against a wall. The bottom of the ladder is 7 feet from the wall. How high up the wall is the ladder?" You have 15 seconds to answer.
>
> "7.14 feet, of course," you reply after a short pause.
>
> "Congratulations, you win a million dollars!"

This scenario would not be a very useful way of demonstrating to students the importance that mathematics might play in their everyday life. Unfortunately, far too many students see mathematics in this absurd way as something that is totally unconnected to anything in their world or in their future world. But what if you could show students how key mathematical concepts play an important role in everyday situations? You most likely would say "great," but be a little wary. After all, you have been bombarded with material that claims to make connections to the "real world," yet it seems that every day in class you answer the question, "When am I ever going to use this stuff?" This book will help. The activities in the book model realistic situations that use mathematics as you do in your life, as a tool or resource for solving problems that may or may not have a set answer. What you won't find are problems that students recognize as contrived and unconvincing, "real-world" problems such as:

> *A garden is in the shape of an obtuse triangle. One angle measures 113 degrees, and the side opposite is 12 feet long. The shortest side is 6.5 feet long. What are the measures of the remaining side and angles?*

What you will find in this book are challenging and engaging activities that have students using geometry to solve meaningful problems and make connections to the world outside the classroom.

How to Use This Book

Organization

This book is organized around six different contexts addressing geometry: Geometry Around Us, Construction and Landscaping, Design and Marketing, Art, Sports and Recreation, and Miscellaneous. For organizational purposes, the topics covered in each activity are listed on the teacher guide page for that activity.

Order of Activities

You'll find that the activities in this book parallel most topics taught in a typical geometry course. You can supplement or enrich a concept presented in your textbook with this resource or use the activities as an introduction to a new concept.

Level of Difficulty

Some activities use more difficult mathematical concepts than others. It should be noted that the less difficult lessons, mathematically speaking, still require higher-order thinking skills.

Time Considerations

Because students' ability levels and schools' schedules vary, time suggestions for the activities are not given. Before using an activity, review it and decide how much time would be appropriate for your students.

Calculators and Other Technology

A practical way of using calculators with the activities is to consider whether or not the situation described in the activity would warrant the use of a calculator in real life. If the situation does, then allow students to use calculators; if it doesn't, then don't allow them to use calculators. In some of the activities, students can use spreadsheet, word-processing, and desktop-publishing software as well.

Organizing the Classroom

The Teaching Notes sections include suggestions on how to arrange students for the activities. Some of the activities work best for individual student work, others are more appropriate for students working in pairs, and some work best for groups of students.

Evaluation and Assessment

Where appropriate, selected answers are given. However, because the lessons model real-life situations, exact answers cannot always be provided.

Teacher Guide Page

1. Talking Geometry

Context

Geometry around us

Topics

Geometric terms

Overview

In this activity, students build confidence about their prior knowledge of geometry by brainstorming about the geometric terms they already know. Then they use those words to create sentences showing how those geometric terms are used in everyday speech.

Objectives

Students will

- list common geometric terms used in everyday speech.
- build confidence in their prior knowledge of geometry.
- create sentences using common geometric terms.

Materials

- Dictionary (optional)

Teaching Notes

1. Students should work in small groups for this activity.
2. This activity is intended as an introductory activity to the class and works best at the beginning of the term. However, it can still be used after the course has started.
3. After all the groups have generated lists of geometric terms, have them turn the lists into you and keep a master list with a frequency tally of terms on the board or overhead.
4. Let students know that they don't have to be able to define each term; they'll do that in Activity 2, Still Talking Geometry.
5. Once all the geometric terms have been posted, have students analyze the list and determine which terms appear most often. There might also be some debate about whether or not a term is actually related to geometry.
6. If students are struggling with creating their sentences, share some sentences that other groups of students have written as they are working.
7. Students do not have to use the classes' most common words when creating their sentences.
8. When students are finished writing their sentences, post the most creative sentences around the room.

(continued)

Teacher Guide Page

1. Talking Geometry *(continued)*

Selected Answers

Some common responses include:

altitude
angle
area
center
circle
circumference
cone
cylinder
degree
diagonal
distance
exterior
height
interior
intersect
length
line
parallel
perimeter
prism
pyramid
rectangle
rotation
square
trapezoid
triangle
volume

Extension Activity

Over the next week, have students listen for geometric terms that they hear in everyday speech and have them keep a list of each word. They will have a chance to use that list in Activity 2, Still Talking Geometry.

Name ______________________________ Date ______________

1. Talking Geometry

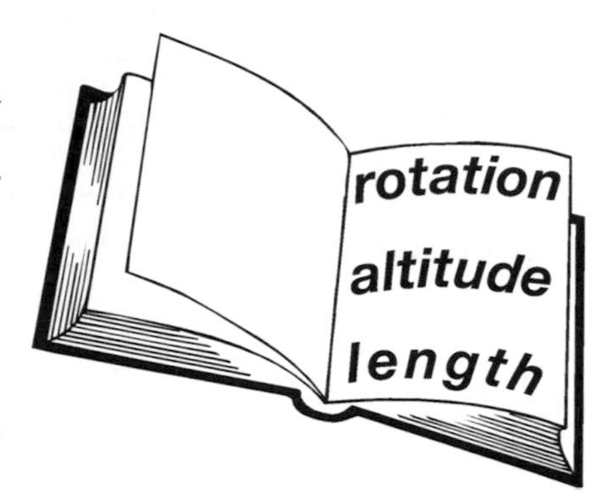

Are you worried that you don't know anything about geometry? You probably know quite a bit more about it than you think. Try this. Work in a small group to think of all the words you already know that are connected to geometry: for instance, *parallel, triangle, circle,* etc., are all words or terms connected to geometry.

List of Terms

List all the words your group comes up with in the space below. If you need more space, use the back of the page.

1. Geometric terms our group thought of ______________________________

   ____

   ____

   ____

2. Give your list of words to your teacher.
3. After your teacher has posted all the words on the overhead or board, list the four most common geometric terms that the class came up with.

   __________ __________ __________ __________

4. In what ways are you surprised, or not surprised, by the list of words?

   ____

   ____

   ____

   ____

Making Sentences

5. In your group, choose any five of the geometric terms from the whole class list and write five sentences using those words. Each sentence you write should use one geometric term in a context or situation where you would naturally use it outside of the classroom. For example, you might say, "In my house there are five windows shaped like rectangles, and two shaped like circles." Of course, the sentences you come up with will be much better than that! Now you try. Underline or highlight the geometric term in each sentence.

(continued)

Name ______________________________ Date ______________

1. Talking Geometry *(continued)*

Sentence 1: ____

____

____

Sentence 2: ____

____

____

Sentence 3: ____

____

____

Sentence 4: ____

____

____

Sentence 5: ____

____

____

6. Pick the best sentence your group wrote out of the five and give it to your teacher.

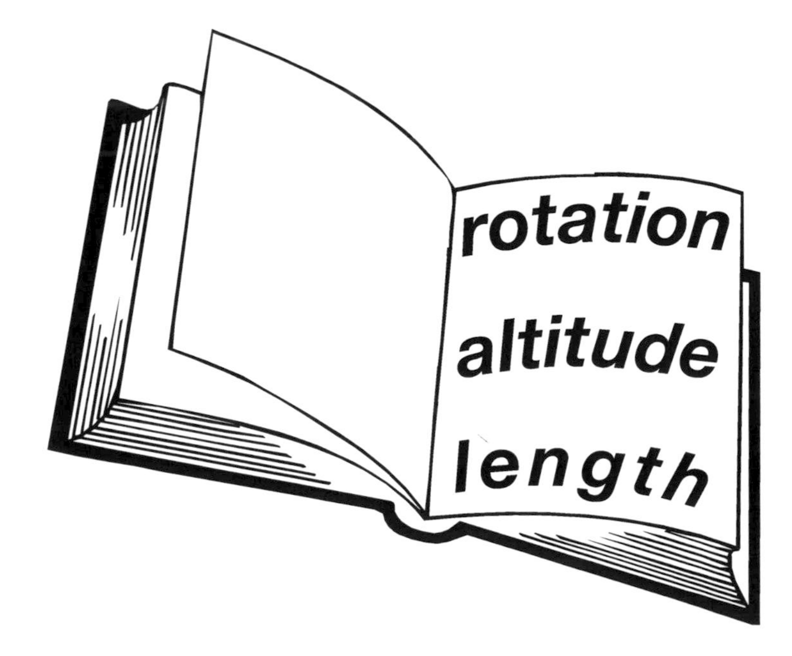

Teacher Guide Page

2. Still Talking Geometry

Context

Geometry around us

Topics

Geometric terms

Overview

In this activity, students make a list of geometry words or terms that they hear, see, or read over a period of time, then use those lists to create a concept map.

Objectives

Students will

- compile a list of geometric terms or words that they hear, see, or read.
- create a concept map using the list of words.

Materials

Access to computers with word-processing or desktop-publishing software (optional)

Teaching Notes

1. Students should work individually to collect their words, and then in small groups to make the master list of words and develop the concept maps.
2. It is recommended that students complete Activity 1, Talking Geometry, before starting this activity.
3. This activity can be used over different lengths of time. Typically, you will want students to create their first concept map after about a month of collecting words. After that, you may have them continue the activity throughout the year by building on their concept maps, or you may stop the activity after the first month.
4. After you assign this activity, verify students' progress in compiling lists of words at least weekly.
5. At the end of the collection period, have students form small groups.
6. It might be necessary to introduce or review the use of concept maps with students before using the lesson.
7. Some students will struggle with finding connections for organizing their concept maps. Suggest ways to find connections; for instance, they can arrange them by geometric topic, description, location, context, and so on.

Selected Answers

Answers will vary depending on the words that students hear, see, and read.

Extension Activity

Have groups put their concept maps on poster paper and allow them to continue the concept map for the entire school year.

Name ______________________ Date ______________

2. Still Talking Geometry

In this activity keep a running list of all the geometry-related words you hear, see, or read outside the geometry classroom.

Create Your List

In the space below, keep a record of any geometry-related words that you hear, read, or see. These words can be from a conversation with a friend, on the radio, on television, in a magazine or newspaper—really, anywhere at all.

Geometry words I heard, read, or saw:

____

____

____

____

____

____

Group List

In your small group, share your list with the other group members and compile one master list.

What similarities and differences did you observe about your list compared with the other students' lists?

____

____

____

____

____

____

(continued)

Name ______________________________ Date ______________

2. Still Talking Geometry *(continued)*

Concept Map

Create a concept map on a sheet of paper using the master list of words. The concept map can be organized in a number of different ways. It is up to your group to establish the relationship between the words.

Explain how the layout of your group's concept map was determined.

____

____

____

____

____

____

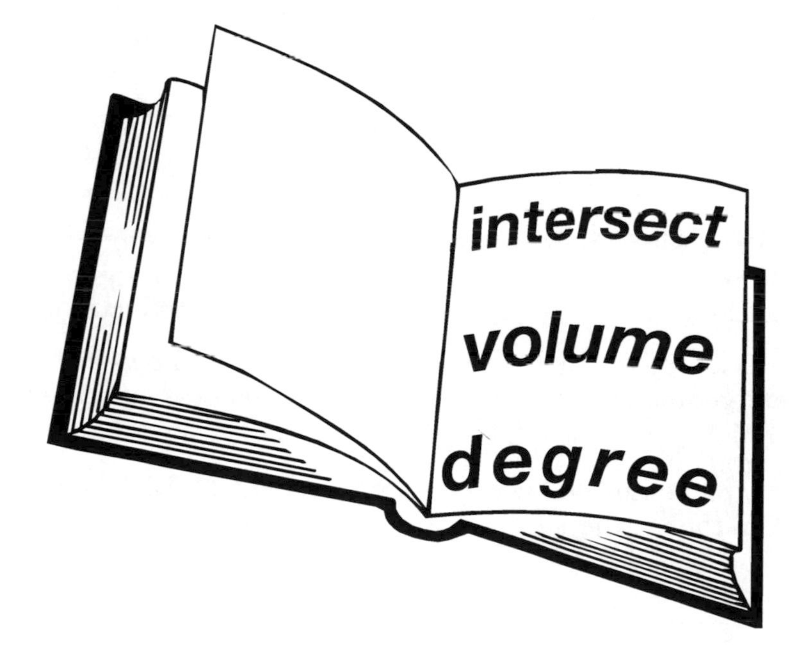

Teacher Guide Page

3. Natural Geometry

Context

Geometry around us

Topics

Underlying geometric principles

Overview

In this activity, students brainstorm ideas about where they might find representations of geometric principles in nature, then they will take pictures of these objects and use them to develop a slide show presentation.

Objectives

Students will

- brainstorm ideas about where they might find representations of geometry in nature.
- take pictures of natural objects that contain or represent geometric principles.
- develop a slide-show presentation using the pictures and data they gathered.

Materials

- Cameras (They can be any type, although digital work best. If only one camera is available, rotate its use among the groups.)
- Access to computer with presentation software

Teaching Notes

1. Students should work in small groups for this activity.
2. This activity can be spaced out over the course of a few classes or it can be used as an ongoing project throughout the year. Availability of natural resources, cameras, and computers will determine length of the activity.
3. Students should not be restricted in what they collect as evidence as long as it occurs naturally and is not man-made. For example, buildings and structures would not qualify, but plants in a garden would be acceptable.
4. Before students begin to develop their slide shows, verify their data in the table on the handout.
5. Some students may not be familiar with using presentation software and may need some help getting started.
6. Once slide shows are completed, have students present them to the class.

Selected Answers

Answers will vary depending on the objects students find.

Extension Activity

Have students investigate how the golden ratio and the Fibonacci sequence occur in nature.

Name ______________________ Date ______________

3. Natural Geometry

The designs on a butterfly's wings, seashells, flowers, rocks and minerals, snowflakes, spiderwebs, vegetables, and plants—these are just a few of the many ways that geometry occurs in nature. What are some other examples of geometric principles that can be found in nature? Brainstorm with your group, and develop a list of possibilities to investigate further.

Possible Examples of Geometry in Nature

List your group's ideas.

____

____

____

Gather Evidence

Find and photograph examples that show underlying geometric principles found in nature. Be sure to document the location, object, and geometric principle for each picture you take. Find at least ten examples.

Geometry in Nature

Object	Date	Location	Geometric Principle(s)

Presentation

Use the photographs you took to develop a presentation documenting what you have found.

Teacher Guide Page

4. Unnatural Geometry

Context

Geometry around us

Topics

Underlying geometric principles

Overview

In this activity, students brainstorm ideas about where they might find man-made representations of geometric principles. They then take pictures of these objects and use the photographs to develop a slide show presentation.

Objectives

Students will

- brainstorm ideas about where they might find representations of geometry around them.
- take pictures of those objects, structures, buildings, or products that contain or represent geometric principles.
- develop a slide show presentation using the pictures and data they gathered.

Materials

- Cameras (They can be any type, although digital work best. If only one is available rotate it among the groups.)
- Access to computer with presentation software

Teaching Notes

1. Students should work in small groups for this activity.
2. This activity can be used as a follow-up to Natural Geometry, in place of Natural Geometry, or in conjunction with Natural Geometry.
3. This activity can be spaced out over the course of a few classes or it can be used as an ongoing project throughout the year. Availability of resources, cameras, and computers will determine length of the activity.
4. Students should not be restricted in what they collect as evidence as long as it is man-made.
5. Before students begin to develop their slide shows, verify their data in the Geometry Around You table.
6. Some students may not be familiar with using presentation software and may need some help getting started.
7. Once slide shows are completed, have students present them to the class.

Selected Answers

Answers will vary depending on the objects students find.

Extension Activity

This activity can be modified or extended with the use of video cameras.

Name ______________________________ Date ______________

4. Unnatural Geometry

Perhaps the opposite of natural geometry is unnatural geometry. In other words, if natural geometry principles can be found in nature, then structures, buildings, products, and so on that are man-made may contain underlying geometric principles. What are some examples of geometric principles that can be found around you? Brainstorm with your group, and develop a list of possibilities to investigate further.

Possible Examples of Geometry Around You

List your group's ideas here.

____

____

____

Gather Evidence

Spend some time photographing examples you find that show underlying geometric principles around you. Be sure to document the location, object, and geometric principle for each picture you take. Find at least ten examples.

Geometry Around You

Object	Date	Location	Geometric Principle(s)

Presentation

Use the photographs you took to develop a presentation documenting what you have found.

Teacher Guide Page

5. Community Service

Context

Construction and landscaping

Topics

Measurement, surface area, estimation

Overview

In this activity, students prepare for a community-service project where they will help refurbish their classroom.

Objectives

Students will

- determine the surface area of a given space.
- develop a cost estimation plan.
- calculate the cost to paint a given area.

Materials

- Calculator
- Tape measure
- Yardstick or meterstick

Teaching Notes

1. Students can work in pairs or small groups for this activity.
2. During the activity, focus students' attention on developing the plan for determining cost—not just on conducting the measurements.
3. For question number one, students will generate ideas such as measuring the room, how to deal with areas that are not painted, cost of paint, cost of brushes and other painting tools, how much paint is needed, and so on.
4. After students brainstorm about what they need to complete the project, let them know that a gallon of paint costs about $15 and covers approximately 300 square feet. These amounts will vary depending on the brand of paint. It may be useful to gather hardware store flyers and let students look through those to determine paint costs.

Selected Answers

Making a Plan

1. Possible answers include: measure the surface area of the room, measure the surface area of surfaces not to be painted, amount of paint needed, cost of paint, cost of brushes, and cost of painting supplies like tape and drop cloths.

2–4. Answers will vary depending on the group's plan and the size of the room.

Extension Activity

Have students make cost estimates for painting all the classrooms at the school.

Name ______________________________ Date ______________

5. Community Service

The school district has asked for volunteers to help refurbish some of the classrooms at your school. You and some of your civic-minded friends have volunteered to assist with the project. As luck would have it, the first project is to paint your classroom. Since the district is on a very tight budget, before you dive in and start painting, you have to plan the job carefully so that costs are kept to a minimum. Follow the steps below to help guide you in developing a plan for determining the cost of painting the classroom.

Making a Plan

1. As a group, consider for a moment what information you might need to ensure that you arrive at the most accurate cost estimate. In the space below, list the ideas your group came up with.

   ____

   ____

   ____

   ____

   ____

2. Compare your ideas with those of one or two other groups. Add anything that you might have left off your list. Write the modifications to your list below.

   ____

   ____

   ____

   ____

3. On the page titled Cost Estimation Plan, organize the group's ideas into a step-by-step list in the proper sequential order. As you organize your plan, keep in mind that the goal, or outcome of the plan, is to be able to make a cost estimate for painting the classroom. Determine the surface area covered by a gallon of paint. Decide how many coats of paint will be necessary.

4. Carry out your plan and organize the data in the Cost Estimation Table. List the cost estimate below.

   Cost estimate: ______________

(continued)

Name ______________________________ Date ______________

5. Community Service *(continued)*

Cost Estimation Plan

___

___

___

___

___

___

___

___

(continued)

Name ______________________________ Date ____________________

5. Community Service *(continued)*

Cost Estimation Table

Location of Wall or Name of Space to Be Painted	**Measurements of Wall and Its Surface Area**	**Measurements of Items Not to Be Painted and Their Surface Areas**	**Surface Area to Be Painted** (subtract figures in column 3 from those in column 2)
	Surface area: ______	Surface area: ______	________
	Surface area: ______	Surface area: ______	________
	Surface area: ______	Surface area: ______	________
	Surface area: ______	Surface area: ______	________
	Surface area: ______	Surface area: ______	________
	Surface area: ______	Surface area: ______	________
	Surface area: ______	Surface area: ______	________
		Total surface area to be painted:	

Teacher Guide Page

6. Landscape Design I

Context

Construction and landscaping

Topics

Area, estimation, scale drawings, unit conversion, percentages, measurement

Overview

In this activity, students assume the role of a landscape designer and determine the amount of sod and its cost to cover an irregularly shaped patch of ground.

Objectives

Students will

- use estimation to measure the area of an irregularly shaped region.
- calculate the area of an irregularly shaped region.
- make scale drawings.

Materials

- Tape measures
- Calculators
- Grid paper (2 pieces)
- Straightedge or ruler
- Protractor (1 per group)

Teaching Notes

1. Students should work in pairs or small groups for this activity.
2. Before the lesson, scout out a suitable region at the school that students can use to measure for sod. Regions should be at least 2000 square feet in area and irregularly shaped.
3. Depending on the number of students in the class and the regions available, you can have each group measure the same region, or you can split up the groups and assign some to another region.
4. Tell students that they will be making two drawings of the region. The first drawing will be a rough sketch with the dimensions of the region recorded on it. They will then use that information to make a scale drawing. The scale drawing can then be used to calculate the area of the region.
5. To improve accuracy, approve each group's sketch before they make their scale drawings.

(continued)

6. Landscape Design I *(continued)*

6. Some students may struggle with determining the area of an irregularly shaped region. Suggest to them that it might be helpful to divide the region into smaller, familiar shapes such as rectangles, circles, or triangles, then calculate those areas individually and add the areas together. Another approach might be to combine different parts of the region together to form familiar shapes.
7. Collect each group's bid amount and compare the results with the whole class.
8. Have students keep their work for the next activity, Landscape Design II.

Selected Answers

Answers will vary depending on the size of the region chosen.

Extension Activity

Have students draft a formal bid using word-processing or desktop-publishing software.

Name ______________________________ Date ______________

6. Landscape Design I

At this point in your geometry career you most likely have had an opportunity to calculate the area for various types of polygons such as quadrilaterals, pentagons, octagons, and triangles. But, in many instances, people have to determine the area of regions that do not fit neatly into a particular classification. In this activity, you and your group will assume the role of a landscape designer competing for a bid to lay sod at your school. To complete the bid, your group will need to apply their knowledge of finding areas for polygons so that they can determine how much sod will be needed and the cost for a given region at the school. In order to submit a competitive bid, your group will not only have to determine the costs for putting in the sod, but also take into consideration that you need to realize a profit. Good luck with the bid process.

Follow the steps below to complete your bid.

Finding Area

1. After your teacher has assigned your group the region where you will lay sod, your group will need to measure it. To help make your group's bid more accurate, you will want to make a sketch of the region. On a sheet of grid paper, make a sketch of the region and accurately record the dimensions. You don't have to worry about the accuracy of the drawing at this point. When you have finished taking the measurements and drawing the sketch, have your teacher look it over.

   Teacher's initials: __________

   After your teacher approves the group's sketch, move on to question two.

2. Use the sketch you made of the region to create a scale drawing on a second sheet of grid paper.

   Explain how your group chose the scale for the drawing.

   ____

   ____

   ____

3. Now that you have your drawing, explain how you will calculate the area.

   ____

   ____

(continued)

Name ______________________________ Date ______________

6. Landscape Design I *(continued)*

4. Calculate the area of your region.

   Area of your region: ______________________________

   Why is the unit of measurement squared?

   ____

   ____

Determining Cost

> *Sod is priced per pallet; one pallet covers 450 square feet and costs $95.*

Step 1. Calculate the cost of the sod, taking into account the cost and the total area to be covered.

Cost of sod: ______________________________

Step 2. Sod is just one of the costs you have to consider. You also have to factor in labor costs. To do this, assume that it will take two workers about one hour to lay sod over 1500 square feet. Workers are paid $11 per hour.

Direct labor costs: ______________________________

Step 3. You will also have to take into account other indirect or overhead costs such as advertising, employee benefits, taxes, and transportation, to name a few. Add 12% to the sod and labor costs for indirect costs.

Indirect costs: ______________________________

Step 4. Keep in mind that you have to make money on the deal. Your company normally adds an additional 8% profit margin.

Amount of profit margin: ______________________________

Step 5. Add the amounts from steps 1–4 together to arrive at the bid amount.

Bid amount: ______________________________

Turn the bid amount in to the teacher for consideration.

Teacher Guide Page

7. Landscape Design II

Context

Construction and landscaping

Topics

Area, estimation, scale drawings, sectors, literal equations, percentages, angle measurements, radius, measurement

Overview

In this activity, students build on the information they learned in Landscape Design I and continue the investigation by adding sprinklers to the bid.

Objectives

Students will

- solve problems and apply solutions to problems involving area.
- solve problems and apply solutions to problems involving sectors of circles.
- determine most cost-effective solution.

Materials

- Scale drawing from Activity 6, Landscape Design I
- Calculators
- Compass (one per group)
- Straightedge or ruler
- Protractor (1 per group)

Teaching Notes

1. Students can work in pairs or small groups for this lesson. You may want them to work in the same groups they were in for Landscape Design I.
2. Before students begin their designs, review with them the criteria (outlined in the handout) that they must follow. It might be useful to demonstrate a partial design on the overhead or board.
3. Pay particular attention to the Additional Criteria section of the activity. Each sprinkler has to throw water to the next sprinkler; this is known as *head-to-head coverage*. Students can divide their sprinkler systems into as many as five zones.
4. Each zone has to have a 0.5-inch precipitation rate per hour. The formula for precipitation rate is given in the handout. Students most likely will have to use a guess-and-check strategy to determine the types and placement of sprinklers to achieve this rate.
5. Remind students that coverage is important, but so is keeping costs down. Remember, the lowest bid that meets specifications generally wins.
6. The last step in the design is crucial. Ensure that students use a compass and protractor to verify coverage zones.

(continued)

7. Landscape Design II *(continued)*

Selected Answers

Answers will vary depending on the size of the region at your school and how students design their sprinkler systems.

Extension Activity

Have students research careers in landscape architecture.

Name ______________________________ Date ______________

7. Landscape Design II

In Landscape Design I, you and your team put together a bid to lay sod over a patch of land at your school. As it turns out, additional money has been allocated for the project, which now makes it possible to put in sprinklers. Time to update that bid! Use the scale drawing you made in Landscape Design I and follow the directions below to put together a bid that now includes a sprinkler system.

Sprinkler Information and Criteria

Use the following information to design your sprinkler system.

- Spray coverage for sprinklers can be adjusted to cover certain sectors, ranging from 90 degrees to 360 degrees.
- Sprinklers are designed to cover certain ranges or distances; these are listed in the table as "radius."
- The number of gallons used per minute (GPM) depends on the sector coverage and range of each sprinkler.
- Cost per sprinkler is in dollars.
- The area watered by each sprinkler must overlap the area watered by the adjacent sprinkler. In other words, the water each sprinkler throws or sprays must reach all the way to the next sprinkler. This is known as *head-to-head coverage.*
- There is sufficient water pressure at the school to support up to five separate sprinkler zones.
- Each sprinkler zone requires an average precipitation rate (amount of water applied to the grass) of 0.5 inch per hour. (See precipitation rate formula on page 23.)
- To save costs, use the minimum number of sprinklers that still meets all the coverage requirements.

As you design your sprinkler system, you can choose any combination of sprinkler types, but you must meet all the criteria listed. To keep your bid competitive, you need to make every effort to keep costs to a minimum.

(continued)

Name ______________________ Date ______________

7. Landscape Design II *(continued)*

Sprinkler Specifications

Sprinkler	Coverage Zone (degrees)	Radius (feet)	Gallons per Minute (GPM)	Cost per Sprinkler
A	90°	5	0.8	$2.00
B	90°	10	1.2	$2.50
C	90°	15	2.0	$3.00
D	180°	5	1.6	$3.25
E	180°	10	2.4	$4.00
F	180°	15	4.0	$4.50
G	270°	5	2.1	$4.75
H	270°	10	2.9	$5.25
I	270°	15	4.5	$6.00
J	360°	5	4.2	$6.25
K	360°	10	6.1	$7.00
L	360°	15	7.9	$7.50

Formula for precipitation rate

$$P_r = \frac{96.25 \times Total\ GPM}{Area}$$

P_r *is precipitation rate in inches per hour.*

96.25 *is a constant that converts gallons per minute to inches per hour.*

GPM *is the cumulative flow from all sprinklers in the specified zone.*

Area *is in square feet.*

Name ______________________________ Date ______________

7. Landscape Design II *(continued)*

Sprinkler Design

Step 1: Divide your region into as many large rectangles or squares as you can. Place sprinklers around the perimeter of each figure, making sure that you have head-to-head coverage and are maintaining a 0.5-inch-per-hour precipitation rate. Along some perimeters it might work best to use 360° sprinklers.

Step 2: Place sprinklers in the remaining areas. Again, remember to have head-to-head coverage and maintain 0.5-inch-per-hour precipitation rate.

Step 3: On your scale drawing, use a compass to show that the coverage area for each sprinkler provides head-to-head coverage.

Costs

1. To find the cost of the sprinklers, multiply the number of each type of sprinkler used by the price for that sprinkler type.

   Cost of sprinklers: ______________________________

2. Piping costs $.79 per foot. For every 10 feet of pipe, a valve is needed. Valves cost $2.40 each.

   Cost of piping and valves: ______________________________

3. As with the sod, take into account the cost of labor. Assume that, for every 10 feet of pipe, labor will cost $22.

   Cost of labor: ______________________________

4. You will also have to take into account other indirect or overhead costs such as advertising, employee benefits, taxes, and transportation, to name a few. Add 12% to the labor and other costs for indirect costs.

   Indirect costs: ______________________________

5. Also, keep in mind that you have to make money on the deal. Your company normally adds an additional 8% profit margin.

   Profit margin: ______________________________

6. Add the total from questions 1–5 together to arrive at the sprinkler system cost amount.

   Cost of sprinkler system: ______________________________

(continued)

Name ______________________________ Date ______________

7. Landscape Design II *(continued)*

Total Bid

Add the bid amount for the sod to the cost of the sprinkler system to arrive at the total bid amount.

Total bid amount: ______________________________

Explain why your bid might be different from the bids prepared by other groups.

____

____

____

____

____

____

Teacher Guide Page

8. The Size of Things to Come, Part I

Context

Design and marketing

Topics

Similarity, surface area, volume

Overview

Student teams assume the role of a marketing group charged with designing a new shape for a soda can. In the first part of this two-part activity, students apply their understanding of surface area to develop a plan for designing a shape for a new soda can.

Objectives

Students will

- generalize how surface area is determined.
- calculate the surface area of an ordinary soda can.
- explain how changing the parameters of a right circular cylinder (soda can) affects its surface area.

Materials

- Soda cans
- Tape measure (one per group)
- Calculators

Teaching Notes

1. Students should work in groups of 3–4 for this activity.
2. The Size of Things to Come, Part I, should be completed before students address The Size of Things to Come, Part II.
3. Set up the activity by eliciting responses to questions such as, "Why are all soda cans basically the same shape?" A follow-up question might be, "How might changing the shape of a soda can influence buyers?" It is not necessary to reach any conclusions to these questions, but they should allow students an opportunity to start pondering the possibility that soda cans might be shaped differently.
4. Students will want to dive right in and begin designing their new soda can shape. It is important that they work through the questions so that they develop a thorough understanding of surface area, and how changing parameters affects the surface area.
5. Let students know that in the follow-up activity, The Size of Things to Come, Part II, they will actually design and construct their new soda can shapes.

(continued)

8. The Size of Things to Come, Part I *(continued)*

Selected Answers

Surface Area

1. Students should make the distinction that surface area is the sum of the existing areas of the faces of the figure.
2. In general, answers should point out that surface area is basically twice the area of the base plus the area of the side or sides.
3. Right circular cylinder.
4. Two circles, a rectangle.
5. The circumference of the base × height + twice the area of the base. The surface area of an ordinary soda can is approximately 49 square inches.

Extension Activity

Next time students are in a store, have them think about the different types, shapes, sizes, and so on of the containers or packages that items come in.

Name ______________________________ Date ______________

8. The Size of Things to Come, Part I

Congratulations! You and your marketing team have just been selected by the New Cola Company to design a new soda can to help launch their new product—Super Duper Cola. In designing the can, you have been given several stipulations. First, like an ordinary soda can, it must hold 12 ounces of soda. Second, it must be made of aluminum. And finally, so that costs are kept to a minimum, the surface area of the can must not exceed the surface area of an ordinary can. Also, it is very important to the New Cola Company that the design is distinctive, that it stands out in a crowd. Good luck with your design!

Surface Area

1. As a group, think about what you already know about area and surface area. As best you can, explain the difference between *area* and *surface area*.

   ____

   ____

   ____

2. Based on what you know about surface area and area, describe a general rule that could be used to calculate the surface area of a solid figure such as a soda can.

   ____

   ____

   ____

   ____

   Compare your group's response with that of another group. Then go on to the next question.

3. To figure out the surface area of an ordinary soda can, a soda can most closely resembles what geometric shape?

   ____

   Measure the soda can. Height: ____________ Diameter: ____________

(continued)

Name ______________________________ Date ____________________

8. The Size of Things to Come, Part I *(continued)*

4. If you could separate a soda can into individual parts and lay each part flat on the table, what different geometric figures would the soda can be composed of?

   ____

   ____

5. Based on your response to question 4, what would be the general formula for determining the surface area of a soda can?

   ____

   ____

   Now, use the formula below to calculate the surface area of the soda can.

   Surface area of a cylinder

   $$S = 2\pi rh + 2\pi r^2$$

   where S = surface area, r = radius, h = height.

   Surface area of ordinary soda can: ____________

6. Without changing the radius, how would **doubling the height** of the soda can affect the surface area?

   ____

   ____

7. Without changing the height, how would **doubling the radius** of the soda can affect the surface area?

   ____

   ____

8. Without changing the radius, how would **reducing the height by half** affect the surface area?

   ____

   ____

(continued)

Name ______________________________ Date ______________

8. The Size of Things to Come, Part I *(continued)*

9. How would **reducing the radius by half** and **doubling the height** of the soda can affect the surface area?

___

___

10. What other questions might be important to consider?

___

___

___

___

___

Teacher Guide Page

9. The Size of Things to Come, Part II

Context

Design and marketing

Topics

Similarity, surface area, ratio, volume

Overview

In this follow up-activity to The Size of Things to Come, Part I, students pick up where they left off as members of a marketing design team developing a new soda can design for the New Cola Company.

Objectives

Students will

- make generalizations about how volume is determined.
- explain how changing the parameters of right circular cylinders (soda cans) affects volume.
- calculate the volume of an ordinary soda can.
- compare the relationship between volume and surface area.
- design a soda can that holds 12 ounces of soda and has a surface area less than or equal to the surface area of an ordinary soda can.
- create a marketing proposal.

Materials

- Soda cans
- Tape measure (one per group)
- Calculators
- Card stock or heavy weight paper
- Word-processing or design software and color printer
- Scissors (one per group)
- Tape or glue (one per group)
- Presentation software (optional)

Teaching Notes

1. Students should stay in the same groups as for The Size of Things to Come, Part I.
2. As in the previous activity, answers should be in terms of π.
3. Again, students will be eager to begin designing the new soda can. Ensure that they work through the questions dealing with volume before creating their designs.
4. Encourage groups to draw designs by hand or with graphics software on a computer.
5. Marketing proposals must demonstrate that the design specifications were followed and should convince readers that this design is the best.

(continued)

9. The Size of Things to Come, Part II *(continued)*

Selected Answers

Volume

1. Students should note that the surface area of a figure is the total of all the areas of all its surfaces; volume is the space inside the figure.
2. In general, the volume of a solid figure is the area of the base times the height.
3. Approximately 29.7π.
4. It would double the volume.
5. It would increase it by the radius squared.
6. It would reduce the volume by half.
7. It would reduce the volume by half.
8. Students should point out that changing the radius produces a larger change because that term is squared when finding volume.

Surface Area vs. Volume

1. 1.22

Extension Activity

Have students explore ideas about other types of packages or containers that could be redesigned to make them more attractive to consumers.

Name ______________________ Date ______________

9. The Size of Things to Come, Part II

Continue with your team's plan for designing the New Cola Company's new soda can.

Volume

1. Explain the difference between *surface area* and *volume*.

____

____

____

____

____

2. Describe a general rule that could be used to calculate the volume of a solid figure such as a soda can.

____

____

____

Share your group's response with another group. How are your responses similar, and how are they different?

____

____

____

3. Calculate the volume of an ordinary soda can.

Volume of a cylinder

$V = \pi r^2 h$

where V = volume, r = radius, h = height.

Volume of an ordinary soda can: ____________π

(continued)

Name ______________________________ Date ______________

9. The Size of Things to Come, Part II *(continued)*

4. Without changing the radius, how would **doubling the height** of the soda can affect the volume?

____

____

5. Without changing the height, how would **doubling the radius** of the soda can affect the volume?

____

____

6. How would **reducing the height by half** affect the volume of the soda can?

____

____

7. How would **reducing the radius by half** and **doubling the height** affect the volume of the soda can?

____

____

8. Explain why changing the radius has a more profound effect on the volume of the soda can than changing the height.

____

____

____

9. Briefly summarize how changing radius and height affects the surface area and volume of the soda can.

____

____

____

____

(continued)

Name ______________________________ Date ______________

9. The Size of Things to Come, Part II *(continued)*

Surface Area vs. Volume

1. What is the **ratio** of the surface area to the volume of an ordinary soda can?

   Ratio: ____________

2. Why is knowing the ratio of surface area to volume important? How might it affect your design?

   ____

   ____

   ____

   ____

Design Time

Now that you have explored surface area and volume, and how changing the parameters affects those measurements, you are ready to create your soda can design. It may take several tries to come up with a design that meets the given specifications, so don't be discouraged. When your team has finished creating the design on paper, build a prototype of your design. Evaluate the prototype and decide if you are satisfied with it. If you are, move on to the next section, Proposal; if not, revisit the design and make modifications as necessary.

Proposal

As important as the design is, the proposal can be just as critical. The proposal will be used to convince the New Cola Company that your design is best. Things to consider when putting together your proposal:

- It is important to convince the New Cola Company executives that your design will stand out from the crowd.
- It should be clear that the design conforms to the specifications as stipulated by the New Cola Company.
- The soda can design should be cost-effective.

Now write a proposal that will convince New Cola that your group has come up with the winning design. Use the back of this handout or a separate sheet of paper.

10. Quilts

Context

Art

Topics

Similarity, area, symmetry, translations, rotations, reflections

Overview

In this activity, students learn how geometric principles are used in creating quilts.

Objectives

Students will

- apply principles of symmetry in designing a quilt.
- use translation, reflection, and rotation to design a quilt.
- calculate area of material used in a quilt design.

Materials

- One-inch grid paper
- Books on quilt design or Internet access to Web sites showing quilt designs
- Marking pens or colored pencils
- Poster board

Teaching Notes

1. Students can work individually or in pairs for this activity.
2. Introduce the activity by finding out if any students are familiar with quilt designs or already know how to make quilts.
3. Students who are already familiar with making quilts will still benefit from the activity since they may not be familiar with the geometric principles of quilt design.
4. Many Web sites showcase quilt designs; these can be used in lieu of printed material. Try searching for *quilt blocks, quilt block patterns, quilt designs.*
5. Students may be somewhat overwhelmed as they preview quilt designs in books or on the Internet. Reassure them that quilt designs such as these take years of practice and that they will be able to make nice designs with just a little practice.
6. You may want to review the concepts of symmetry, rotation, reflection, and translation with students before the activity.
7. At first, students unfamiliar with quilting should stick to relatively simple designs using triangles or squares.
8. Once students have created their basic design grid, you will need to make photocopies of the design.

(continued)

Teacher Guide Page

10. Quilts *(continued)*

9. Using copies of their basic design grid, students will experiment with arranging the grids into a pattern or template that would be used to make the quilt. After they have completed the template, students should tape or glue the design together, reproduce it, and color it in as they desire.
10. Students should describe how their pattern works using geometric terms such as translation, rotation, symmetry, and reflection.

Selected Answers

Answers will vary depending on students' designs.

Extension Activity

Students can investigate other ways in which quilts are used besides as bed coverings.

Name ______________________________ Date ____________________

10. Quilts

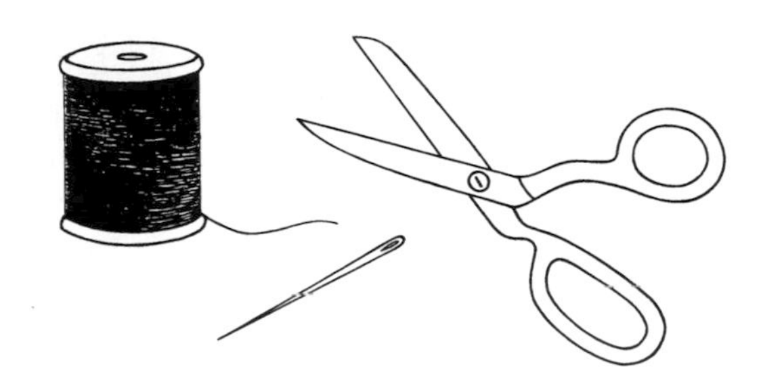

Quilts are a traditional American form of art that have been around for centuries. Quilts are typically constructed by designing a block pattern or template on a grid. Each individual block is then joined with other blocks to form the overall pattern. Follow the steps below to create your own quilt design.

On the Internet, search for a Web site that shows quilt designs. While you are looking over the designs, consider how the patterns work. Pay particular attention to the symmetry, rotation, reflection, or translations that might be present in the design. Pick one quilt pattern to study in greater detail.

Using the quilt pattern you chose, describe its pattern in terms of symmetry, rotation, reflection, or translations. In the grid below, draw the basic design.

____

____

____

____

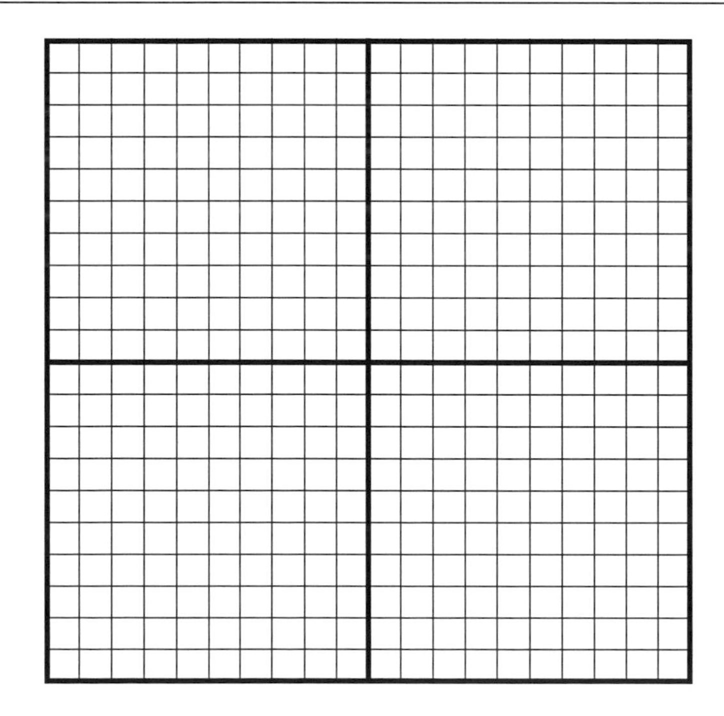

(continued)

Name ______________________________ Date ______________

10. Quilts *(continued)*

Grid Design

Now it is your turn to come up with a design of your own.

Step 1: On a sheet of grid paper, outline a square. The square will form the foundation of your quilt design.

Step 2: Make a pattern inside the square. For now, keep the design simple. For example, you may want to create a design using symmetrical triangles.

Step 3: Draw the reflection of your design.

Step 4: Make 6–8 copies of the design and its reflection.

Step 5: Using the squares you made, create different designs using rotations, reflections, and translations. When you are satisfied with a pattern, tape or glue it onto a piece of poster board.

Write a brief description of how your design works.

____

____

____

Problem Solving with Your Design

1. Suppose you wanted to make your quilt fit a queen-size bed. How much material of each color would you need? A standard-sized queen mattress measures 80×60 inches.

____

____

2. Suppose you wanted to make your quilt to fit on a king-size bed. How much material of each color would you need? A standard-size king mattress measures 84×72 inches.

____

____

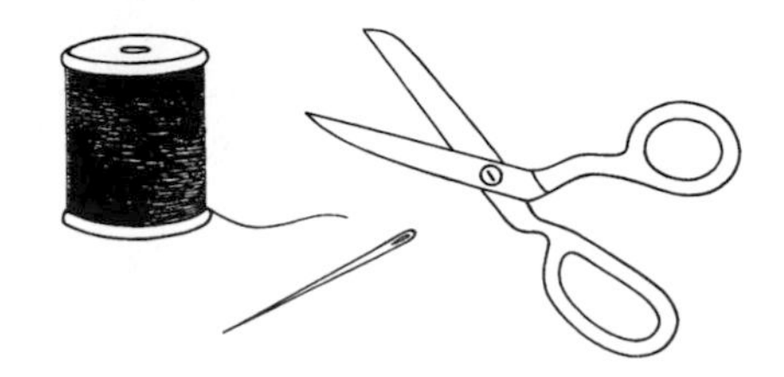

Teacher Guide Page

11. Making Really Big Pictures

Context

Art

Topics

Similarity, scale drawings, measurement

Overview

In this activity, students use similarity to create a mural.

Objectives

Students will

- apply principles of similarity to create a larger picture from a smaller one.
- develop a working definition for similarity.

Materials

- Grid paper
- Butcher paper or poster board
- A small picture, drawing, photograph, or cartoon to enlarge (students can select)
- Straightedge or ruler
- Marker pens or colored pencils

Teaching Notes

1. Students should work individually on this activity.
2. Before using this activity, direct students to look for murals or other large images that they see around town.
3. Students will need to create their own drawing or use an existing design, drawing, or photograph during the activity. Either have students bring in an image to enlarge or direct them to find suitable drawings or designs in magazines, using software, or on the Internet.
4. It is important that students practice enlarging a simple drawing before tackling a more complicated piece. Verify students' work before they move on to the Making Your Mural section of the activity.
5. In choosing a design or drawing, you may want to assign a particular theme or topic, such as nature, current events, social issues, sports, or even geometry.
6. Encourage students to be patient and not try to rush their enlargements. Emphasize how important it is to work on one square at a time.

Selected Answers

Answers will vary.

Extension Activity

Have students research the history of murals in their town. If possible, they might interview the artists of those murals.

Name ______________________ Date ______________

11. Making Really Big Pictures

You have seen them on the sides of buses, covering buildings, at the mall; there may even be one on the side of your school. What are we talking about? Murals, of course! Murals are found all over the world and have been a popular art form for centuries. To create a mural, artists start with a small drawing or design, and use the principle of similarity to enlarge it.

In this activity, you will have a chance to create your own mural design. You can make your own drawing or use an existing picture, cartoon, or drawing. Even if you are artistically challenged, you can still create a mural using your knowledge of geometry.

Think about a mural or large drawing that you have seen recently. Describe how you think the artist went about making such a large drawing. In other words, what are the steps an artist would use to turn a small drawing into a larger one?

____

____

____

____

Share your response with another student.

Practice Drawing

Before you jump into making a mural, and mess up a wall at school or something, let's practice on a simple design so that you have a full understanding of the process.

Get a sheet of grid paper and divide it in half by drawing a line across the middle of the paper. In the top half of the page, draw a simple picture of a house. The house should take up about one quarter of the entire paper. Keep the drawing simple, maybe a door and two windows. Next, using the bottom half of the paper, double the size of the picture you just drew by using a 1-to-2 scale. Each small square in the upper drawing will correspond to a 2×2 square in the lower drawing. To create the drawing, sketch exactly what is in each smaller square into the corresponding larger square. It works best if you only draw one square at a time and then stay within that individual square as you enlarge it.

How did the drawing turn out? If you are satisfied with the results, move on to the next section, Making Your Mural. If you still need more practice, get another piece of grid paper, make another drawing, and try it again. When you are ready, move on to the next section.

(continued)

Name ______________________________ Date ______________

11. Making Really Big Pictures *(continued)*

Making Your Mural

The process for making a mural is about the same as you just practiced. One difference is that the original drawing won't be on grid paper. Therefore, you have to superimpose a grid onto the drawing that you want to enlarge.

Step 1. The first thing you need to do is decide on a design to enlarge, or you can make a drawing yourself and use that.

Step 2. Once you have decided on the drawing or picture that you are going to use, lightly draw a 1-cm grid of squares over it.

Step 3. On the poster paper, draw a similar grid. The size of the grid is determined by the size of the poster paper.

Size of the grid on the poster paper: ______________________________

Step 4. Create the mural by drawing exactly what is in each smaller square into the corresponding larger square. It works best if you work on one square at a time and stay within that individual square. Remember to shade and color as necessary also.

Now that you have worked with the concept of similarity, spend a few minutes thinking about what exactly it means. Then, in your own words, write a definition of similarity.

____

____

____

____

____

____

Teacher Guide Page

12. Line Designs

Context

Art

Topics

Algebra, symmetry, proportion

Overview

In this activity, students learn how symmetry is used in line designs.

Objectives

Students will apply principles of symmetry in creating line designs.

Materials

- Grid paper
- Straightedge

Teaching Notes

1. Students should work individually for this activity.
2. Students generally catch on quickly to how to construct line designs. If some students are having difficulty, you may want to demonstrate on the board or overhead.
3. Encourage students to modify the spacing of the tick marks to create other designs.

Selected Answers

1. Students should note that the straight lines now appear curved.

Extension Activity

Have students make line designs using colored strings.

Name ______________________________ Date ______________

12. Line Designs

One of the most common geometric principles found in art is symmetry. A good example of the use of symmetry in art is in line designs. Follow the steps below to create a line design using symmetry.

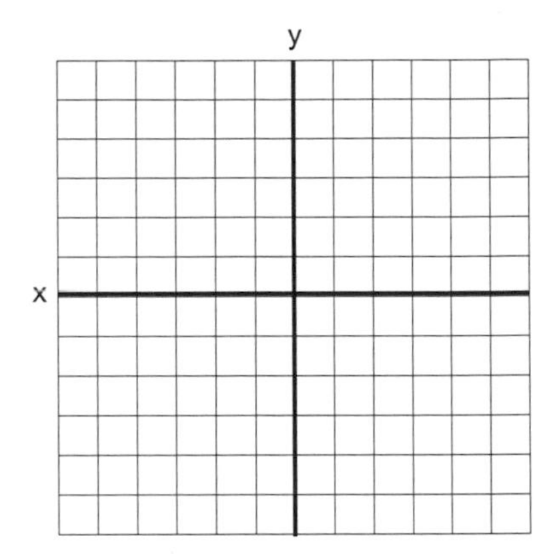

Creating Line Designs

Step 1: Draw an x-axis and a y-axis on a sheet of grid paper. Each axis should be about 10 cm long. Make tick marks every five millimeters on each axis. Label the tick marks outward from the origin 1–10.

Step 2: In quadrant one, draw a straight line from the number 10 tick mark on the y-axis to the number 1 tick mark on the x-axis. Next, draw a line from the number 9 tick mark on the y-axis to the number 2 tick mark on the x-axis. Continue connecting tick marks in this manner until all have been connected.

Step 3: Repeat the process from step 2 in quadrant two.

Step 4: Repeat the process from step 2 in quadrant three.

Step 5: Repeat the process from step 2 in quadrant four.

Observations

1. What do you notice about the straight lines that you have drawn?

____

____

____

2. Describe the symmetry in your design.

____

____

____

Create Your Own Design

Create your own design by modifying the spacing of the tick marks.

Teacher Guide Page

13. Racetrack

Context

Sports and recreation

Topics

Circumference, measurement, diameter, radius

Overview

In this activity, students determine the starting points on a running track for a 400-meter race.

Objectives

Students will

- calculate circumference when given radius.
- calculate total distance of a track lane when given radius and the length of the straightaway.
- design a four-lane oval track with straightaways and semicircular ends.
- determine runners' starting points for a 400-meter race on the running track of their design.

Materials

- Trundle wheel (one per group)
- Calculator
- Tape measure (one per group)

Teaching Notes

1. Students can work in pairs or small groups for this activity.
2. For this activity you will use the running track at your school. If your school does not have a track, you may have to plan on visiting a nearby track, or you may choose to measure a track yourself and provide the measurements for the class.
3. From watching track events at school, or televised events events such as the Olympics, many students have some familiarity with the way certain track events use staggered starting points.
4. Students who participate in track events may have experience using staggered starting points. It can be useful to have them explain why runners are staggered at the beginning of a race.
5. If the starting points are already marked on the track at your school, when students go out to measure the track, do not allow them to measure the premarked starting points.

Selected Answers

Practice Run

1. Students' responses should address the idea that the distance around each lane increases as the lane moves further from the center of the track. As a result, each runner's starting point is moved forward to compensate for the increased length of the lane.

(continued)

Teacher Guide Page

13. Racetrack *(continued)*

2. The radius of the two semicircles, the length of the straightaway, and the width of the lane.

Lane One	Lane Two	Lane Three	Lane Four
$r_1 = 25$	$r_2 = 26$	$r_3 = 27$	$r_4 = 28$
$s = 50$	$s = 50$	$s = 50$	$s = 50$
$t = 2\pi r + 2s$ $t = 50\pi + 100$	$t = 52\pi + 100$	$t = 54\pi + 100$	$t = 56\pi + 100$
	Difference between lane one and lane two: 2π	Difference between lane one and lane three: 4π	Difference between lane one and lane four: 6π

3. The length of each runner's head start.
4. $2\pi r$ represents the distance around the two semicircles of the track, and $2s$ represents the length of the two straightaways.

Your Race Track

1–4. Answers will vary depending on the dimensions of the track at your school.

5. It would not change the runners' starting points since the length of the straightaway is the same for each lane.

Extension Activity

Students can research to determine if starting in a particular lane gives a runner any advantage in a race.

Name ______________________________ Date ______________

13. Racetrack

Congratulations on your new job as the high school track coach! I know you have been busy recruiting members, organizing practices, and scouting the competition. Unfortunately, you have overlooked one little thing: It seems that the runners' starting places on the track are not marked and you have forgotten to determine the starting points for the upcoming track meet. Good thing you paid attention in geometry class so that you can quickly determine those starting points. Okay, so maybe you did not pay as much attention as you should have, or maybe you kept asking yourself, "When am I ever going to use this stuff?" Regardless, with a little brushing up, you will be able to figure out those starting points in no time.

Practice Run

1. Why is it necessary for runners in a race such as an 800-meter race to use staggered starting points?

____

____

____

2. What information do you need to know in order to figure out the head start for each lane?

____

____

The track pictured below has an inner radius of 25 m, straightaways that measure 50 m, and a constant lane width of 1 m. Use that information to fill out the table on the next page.

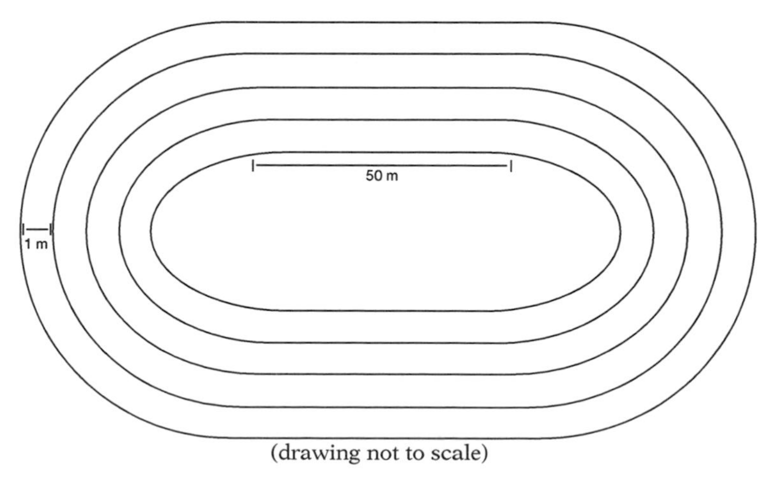

(drawing not to scale)

(continued)

Name ________________________________ Date ________________

13. Racetrack *(continued)*

Sample Track Dimensions

Lane One	Lane Two	Lane Three	Lane Four
$r_1 = 25$	$r_2 =$	$r_3 =$	$r_4 =$
$s = 50$	$s =$	$s =$	$s =$
$t = 2\pi r + 2s$ $t = 50\pi + 100$	$t =$ $t =$	$t =$ $t =$	$t =$ $t =$
	Difference between lane one and lane two:	Difference between lane one and lane three:	Difference between lane one and lane four:
	Let r = radius of inner lane, s = length of a straightaway, t = total distance around a lane.		

3. What does the last row of the table represent?

____

____

____

4. In the equation $t = 2\pi r + 2s$, what does $2\pi r$ represent?

____

What does 2s represent?

____

(continued)

Name ______________________________ Date ______________

13. Racetrack *(continued)*

Your Racetrack

For the next part of the activity, use the track at your school for taking measurements.

1. Find the measurements of the track at your school and fill in the table below.
   (a) Length of the straightaway: ______________________________
   (b) Radius of the inner circle: ______________________________
   (c) Lane width: ______________________________

Track Dimensions

Lane One	Lane Two	Lane Three	Lane Four
$r_1 =$	$r_2 =$	$r_3 =$	$r_4 =$
s =	s =	s =	s =
t = t =	t = t =	t = t =	t = t =
	Difference between lane one and lane two:	Difference between lane one and lane three:	Difference between lane one and lane four:
	Let r = radius of inner lane, s = length of a straightaway, t = total distance around a lane.		

(continued)

Name ______________________________ Date ______________

13. Racetrack *(continued)*

2. In the space below, draw a diagram that represents the track at your school. Include its measurements.

3. To set up for a 400-meter race, where would the starting points for the runners need to be placed? Indicate the starting points on your diagram.
4. To set up for an 800-meter race, where would the starting points for the runners need to be placed? Indicate the starting points on your diagram.
5. If the track at your school were a circle, describe how that would alter the placement of the runners' starting points.

____

____

____

6. Based on the data in tables one and two, what generalizations can you make about the measurements of all tracks?

____

____

Teacher Guide Page

14. Air Supply I

Context

Sports and recreation

Topics

Measurement, volume, literal equations, problem solving, estimation

Overview

In this activity, students learn how divers apply gas laws to help plan scuba dives.

Objectives

Students will use Boyle's Law to determine how change in pressure affects gas volume at constant temperature.

Materials

- Calculator

Teaching Notes

1. Students can work individually, in pairs, or in small groups for this activity.
2. No prior knowledge of scuba diving is needed by the teacher or students to complete this activity.
3. It may be possible to coordinate this activity and the follow-up activity, Air Supply II, with a unit in a science class such as physics.
4. Some students may find it easier to understand how the change in pressure affects volume if they make a drawing depicting change in volume as pressure increases.
5. When working with gas laws, pressure is measured in atmospheres absolute (AtA), which takes into account existing atmospheric pressure of 14.7 psi or 1 atm.

Selected Answers

Boyle's Law

1. 20 cubic feet.
2. 31.43 cubic feet.
3. 18.06 cubic feet.
4. There is insufficient air remaining to allow the dive to be completed.

Extension Activity

Have students consider how the volume of a gas might be affected under conditions of decreasing pressure.

Name ______________________ Date ______________

14. Air Supply I

Divers have to plan for many different things before they actually get in the water. One of the most important things they have to plan for is how much air they will need for a scuba dive. As such, dive supervisors learn how such things as changes in pressure, temperature, and breathing rates affect the available volume of air.

Suppose you are a dive supervisor planning a scuba dive. First, you will explore the relationship between volume and pressure, as described by Boyle's Law. Second, you will apply what you have learned to determine the duration of the air supply for a scuba dive.

Boyle's Law

One of the first gas laws that divers learn about is Boyle's Law. Boyle's Law explains the relationship between pressure and volume: At constant temperature, the absolute pressure and the volume of a given mass of gas are inversely proportional. As you go deeper in the water, the pressure increases, and the volume of the gas decreases; or you could say the opposite—as the pressure decreases, the volume increases. Understanding this relationship is important because as a diver goes deeper, less air is available. For example, suppose you had a one-gallon milk container at the surface of the water. This milk container is under 1 atmosphere (atm) of pressure. One atm is equal to 14.7 pounds per square inch, or psi. If you inverted the milk container and took it to a depth of 33 feet, it would be under 2 atm of pressure (29.4 psi). Every 33 feet of water is equal to 1 atm of pressure. Because there is now twice the absolute pressure on the container, the volume is decreased to one-half gallon. If you took the container to a depth of 66 feet, then the pressure would be equal to 3 atm, and would compress the volume to one-third gallon.

Boyle's Law can be expressed as:

Boyle's Law

$$P_1V_1 = P_2V_2$$

Where P_1 = *initial pressure,*
V_1 = *initial volume,*
P_2 = *final pressure, and*
V_2 = *final volume*

(continued)

Name ______________________________ Date ______________

14. Air Supply I *(continued)*

When working with Boyle's Law, pressure must be measured in atmospheres absolute (P_{ata}), which takes into account the existing 1 atm [14.7 pounds per square inch (psi)] of pressure at sea level. A gauge at sea level that reads zero (0 psig, or zero pounds per square inch gauge) would be expressed as 14.7 psia to register atmospheric pressure. To calculate pressure using atmospheres absolute,

$$P_{ata} = \frac{Depth(ft) + 33ft}{33ft}$$

Follow the steps below to solve for final volume, V_2.

A. Using the formula for Boyle's Law, solve for V_2.

B. Draw a diagram.

C. Calculate the final pressure, P_2, at the depth the diver descends to.

D. Substitute known values back into the formula for Boyle's Law and solve for the final volume, V_2.

1. If a diver starts a dive with a tank filled with 80 cubic feet of air and immediately descends to 99 feet, what is the volume of air in the tank at 99 feet?

   Tank volume: ______________

2. If a diver starts a dive with two tanks filled with 80 cubic feet of air each and immediately descends to 135 feet, what is the volume of air available for the diver at 135 feet?

   Tank volume: ______________

3. If a diver starts a dive with one tank filled with 80 cubic feet of air, descends to 120 feet, uses up half the volume of air, then ascends to 40 feet, what is the volume of air in the tank? Draw a diagram to support your response.

   Tank volume: ______________

(continued)

Name ______________________________ Date ______________________

14. Air Supply I *(continued)*

4. A diver carrying twin 72-cubic-inch tanks descends to 85 feet to inspect a piling. The diver spends 10 minutes at that depth and uses up 20% of the air supply. If the diver needs at least 5 cubic feet of air to inspect the bottom of the piling at 130 feet, explain whether or not the dive can be completed.

____

____

____

____

____

5. Create a diving scenario and have a classmate solve it.

____

____

____

____

____

____

Teacher Guide Page

15. Air Supply II

Context

Sports and recreation

Topics

Measurement, volume, literal equations, problem solving, estimation

Overview

In this activity, students continue learning how divers apply gas laws to help plan scuba dives.

Objectives

Students will

- use tables and formulas to calculate air consumption rates.
- use given formulas to calculate available air capacity.
- use given formulas to calculate the duration of available air supply.

Materials

- Calculator

Teaching Notes

1. Students can work individually, in pairs, or in small groups for this activity.
2. This lesson builds on information on the relationship between pressure and volume that the students learned about in the first part of the activity, Air Supply I.
3. No prior knowledge of scuba diving is needed by the teacher or students to complete this activity.
4. Many students will find it useful to draw diagrams when solving the problems.

Selected Answers

1. 26 minutes
2. Duration of available capacity in minutes: 19. Too little time for a 25-minute security swim of the ship's hull.
3. 18 tanks.

Extension Activity

Have students research how other gas laws are used in diving.

Name ______________________________ Date ____________________

15. Air Supply II

In the first part of the activity, you explored the relationship between pressure and volume. Now you will combine that information with a new formula to determine the length of time that a diver can stay underwater.

Duration of Air Supply

The duration of the air supply depends upon

- The diver's air-consumption rate
- The depth of the dive
- The capacity and recommended minimum pressure of the tank(s)

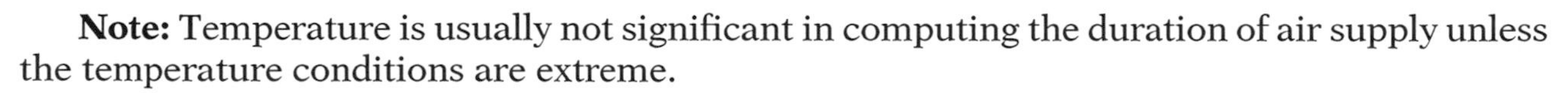

Note: Temperature is usually not significant in computing the duration of air supply unless the temperature conditions are extreme.

There are three steps in calculating how long a diver's air supply will last.

Step 1: Calculate the diver's air-consumption rate by using this formula:

Formula for air-consumption rate

$$C = \frac{D + 33}{33} \times RMV$$

Where: C = diver's air-consumption rate, standard cubic feet per minute (scfm)
D = depth
RMV = diver's respiratory minute volume, actual cubic feet per minute (acfm)

Example: If a diver is doing moderate work at 90 feet, to calculate the diver's air-consumption rate:

$$C = \frac{90 + 33}{33} \times 1.4$$

$$= 5.22 \text{ scfm}$$

(continued)

Name ______________________________ Date ______________

15. Air Supply II *(continued)*

Step 2: Calculate the available air capacity provided by the tanks. The formula for calculating the available air capacity is

Formula for available air capacity

$$V_a = \frac{P_c - P_{rm}}{14.7} \times (FV \times N)$$

Where V_a = capacity available (scf)
P_{rm} = recommended minimum pressure of tank, psig
P_c = measured tank pressure, psig
FV = internal volume of tank (scf)
N = number of tanks

Example: If the same diver were using a set of twin 80-cubic-foot tanks charged to 3000 psig, to calculate available air

$$V_a = \frac{3000 - 500}{14.7} \times (0.399 \times 2)$$

$$V_a = 135.7 \text{ scf}$$

Step 3: Calculate the duration of the available capacity (in minutes) by using this formula:

Formula for duration of available capacity

$$\text{Duration} = \frac{V_a}{C}$$

Where V_a = capacity available (scf)
C = diver's air-consumption rate (scfm)

Example: Using the information from the previous two examples:

$$\text{Duration} = \frac{135.7 \text{ scf}}{5.22 \text{ scfm}}$$

$$\text{Duration} = 26 \text{ minutes}$$

(continued)

Name ______________________________ Date ____________________

15. Air Supply II *(continued)*

Practice

1. Determine the duration of the air supply of a diver doing moderate work at 70 feet using twin 72-cubic-foot steel cylinders charged to 2,250 psig.

   RMV = 1.4 acfm

   FV = 0.420 scf

   P_{rm} = 250 psig

   Duration of air supply: ______________

2. After a long day of diving, you have one last job to complete—a security swim of the ship's hull. From experience, you know that it takes 25 minutes to complete this. You only have one 80-cubic-foot tank left, charged to 3,000 psig. Now you have to decide if you need to go back to the base for additional tanks or if you can complete the job without having to go back in to fill additional tanks.

   D = 40 feet

   RMV = 1.59 acfm

   FV = 0.399 scf

   P_{rm} = 500 psig

   Explain whether or not you can complete the dive without returning to the base.

   ____

   ____

3. You and your dive team have been assigned to recover a missile that fell from an airplane. The information you have indicates that the missile is located in 90 feet of water. You anticipate that you will spend at least two hours locating the missile and another two hours recovering it. How many 80-cubic-foot tanks do you need to bring with you?

   D = 90 feet

   RMV = 1.34 acfm

   FV = 0.399 scf

   P_{rm} = 500 psig

   P_c = 3000 psig

   Number of 80-cubic-foot tanks needed to complete the mission: ______________

4. If the depth of the missile in problem three were only 30 feet, how would that affect the number of tanks required for the dive? Explain your answer.

   ____

   ____

Teacher Guide Page

16. Networks

Context

Miscellaneous

Topics

Logical reasoning, measurement, networks

Overview

In this activity, students use logical reasoning to design a network for collecting attendance records at the school.

Objectives

Students will

- use logical reasoning to design the most efficient route (network) for collecting attendance figures at the school.
- measure the network.

Materials

- Campus map
- Trundle wheel (one per group)
- Calculator

Teaching Notes

1. Students should work in groups of 3–4 for this activity.
2. Depending on the size of your school, you may need to divide the school into smaller zones so that students can complete the activity in a reasonable amount of time.
3. Familiarize yourself with the network design guidelines before having students complete the activity.
4. Depending on the layout of your school, it may be necessary for students to measure only a section of the school and make assumptions about distances for symmetrical sections. If this is possible, elicit from students how this might work.
5. Before students actually go out and measure the network they designed, you may want to review their plans to ensure that they are on the right track.
6. Some students, when measuring their networks, will think in terms of speed rather than distance. Emphasize that speed is treated as a constant for this activity.

Selected Answers

Network Design Procedures

Students should indicate that it would be possible to measure the distance if the campus map were drawn to scale.

Network Evaluation

Answers will vary depending on network designs.

Extension Activity

Students can extend the activity by designing a mail delivery network for their neighborhood.

Name ______________________________ Date ______________

16. Networks

Whenever you travel between several different points, you are traveling on what is known as a **network.** A network is a collection of points connected by paths. Networks are designed and used by the post office, delivery companies, phone companies, cable companies, and others to ensure efficiency when traveling between points.

Networks must be designed to be as efficient as possible. One network that probably exists at your school is the route used to pick up daily attendance figures from the classrooms. However, the current network may not be the most efficient one possible. It is up to your team to design the most efficient network to collect attendance figures from each classroom.

Network Design Guidelines

- The network must start and finish at the school's attendance office.
- Every classroom that normally takes attendance must be included in the network.
- Network paths may cross.
- All attendance takers walk at the same average speed.
- Each network must be measured using a trundle wheel.
- The network with the shortest overall **distance** is the winning design.

Network Design Procedures

Step 1: Mark an X on the campus map at the attendance office. The X will represent the starting and stopping point.

Step 2: Carefully study the layout of the school. As a group, ask and answer such questions as, "What's the most efficient way to get to all the classrooms?" and "How do we minimize the distance that the attendance taker must travel?"

Step 3: Draw your plan on the campus map. Review it carefully, and make any revisions that become apparent. It most likely will take several attempts before you are satisfied with the network design.

(continued)

Name ______________________________ Date ______________

16. Networks *(continued)*

Suppose that you were unable to physically go out and measure the distance of your network design. How else might you measure it?

____

____

____

Step 4: Now you are ready to go out and measure your network design. Using your trundle wheel, start at the attendance office and measure the entire distance of the network. To facilitate the process, have one member of your group record distances and another walk with the trundle wheel; have another keep track of the route.

Step 5: After you have completed measuring the distance of your network, record the total distance below.

Total network distance: ______________

Network Evaluation

1. How would it be possible to shorten the distance of your network?

____

____

____

2. Compare your design with several other groups' designs. What are some of the similarities and differences between the designs? How do the distances differ?

____

____

____

3. In what other settings would it be useful to understand the principles of network design?

____

____

____

4. If speed were taken into consideration, how would the design of the network be affected?

____

____

Teacher Guide Page

17. Flowcharts

Context

Miscellaneous

Topics

Proof, logic, reasoning

Overview

In this activity, students learn how flowcharts are used to make a pictorial representation of a complicated system or process.

Objectives

Students will

- identify and use different symbols used in flowcharts.
- create accurate flowcharts that represent different processes.

Materials

- Paper
- Word-processing or desktop-publishing software (optional)

Teaching Notes

1. Students can work individually, in pairs, or in small groups for this activity.
2. Students may be familiar with using flowcharts from computer classes; however, some students will not have used them. You may need to spend some time introducing how they are used.
3. You may also want to bring in several examples of flowcharts to introduce the lesson. Besides computer applications, flowcharts are commonly used in medical fields, schools, and business. There most likely will be examples of flowcharts in your geometry textbook.
4. When students are ready to make their own flowcharts, have them start with a process that they are very familiar with. For example, they might design a flowchart for a sport, a hobby, or an everyday routine.

Selected Answers

There is no one set way to design a flowchart. However, pay particular attention to the use of the standard symbols; these should reflect their intended purpose.

Extension Activity

Have students research and gather examples of flowcharts that they find.

Name ______________________________ Date ______________

17. Flowcharts

A **flowchart** is a pictorial representation showing a step-by-step procedure through a complicated system or process. A flowchart helps you to visualize a logical thinking process. Flowcharts are used in a variety of ways. For instance, medical decisions can be determined by following a flowchart. In the business world, flowcharts are used in many different ways. In fact, flowcharts can be found in almost every career field. Typically, flowcharts start with a question requiring a decision and use boxes to represent actions; arrows connect the boxes to show how the process flows. In this activity, you will learn how to develop a flowchart to represent a process you are familiar with.

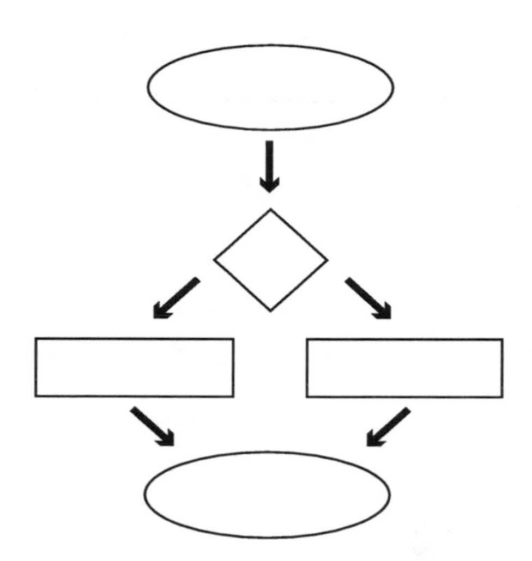

Steps for Creating a Flowchart

Decide on a process or procedure to use.

1. Write the opening question.
2. Determine the major decisions to be made in the process.
3. Consider alternatives for each process.
4. Write an outcome or end to the process.

Other considerations:

- The first symbol in the flowchart represents the beginning of the process and is represented by an **elongated oval.**
- Decision points are represented by **diamonds.**
- Tasks or process blocks are represented by **rectangles.**
- The last symbol in the flowchart represents the end or result of the process; it is represented by an **elongated oval.**

(continued)

Name ______________________________ Date ______________

17. Flowcharts *(continued)*

Create Your First Flowchart

Start by designing a relatively simple flowchart showing the process for taking high school mathematics courses so that you can take calculus.

1. Write the question for the start of the flowchart.

____

____

2. What are the major decisions that have to be made in the process?

____

____

____

3. What are some of the alternatives for each decision?

____

____

____

4. What is the outcome of the process?

____

____

On a separate sheet of paper, use the answers to questions 1–4 above to construct your flowchart.

More Flowchart Practice

On a separate sheet of paper, construct a flowchart to represent the following processes:

A. The steps for solving a word problem in math

B. The steps for solving a linear equation

Make Your Own Flowchart

Create two flowcharts for any processes that you want.

Flowchart topic number one: ______________________

Flowchart topic number two: ______________________

Teacher Guide Page

18. Angled vs. Straight

Context

Miscellaneous

Topics

Area, measurement, estimation, scale drawings

Overview

In this activity, students consider what is required to maximize the available space in a parking lot, then use that information to create a design for a parking lot.

Objectives

Students will

- research local parking lot ordinances to determine parking lot measurement requirements.
- brainstorm ideas about other issues to consider when designing a parking lot.
- design a parking lot that maximizes the number of spaces while at the same time meeting the requirements of the local parking ordinances.

Materials

- Grid paper
- Ruler
- Protractors
- Computers with Internet access (optional)

Teaching Notes

1. Students can work individually, in pairs, or in small groups for this activity.
2. Since there is no one standard parking space measurement, students will have to research local requirements. Much of that information can be found on the Internet or by calling local government offices.
3. Build up the activity by asking students to pay attention to parking lots for a few days before starting the lesson.
4. Have students use grid paper to make their designs.

Selected Answers

Other issues students may consider include requirements for handicapped parking, compact spaces, and landscaping.

Extension Activity

Have students research other examples of how geometric principles are applied in zoning, ordinances, or land-use issues.

Name ______________________________ Date ________________

18. Angled vs. Straight

Over the years, you may have noticed that parking lots are not all designed the same. Some parking lots have angled spaces, and some have straight spaces. Why do you suppose that is? Which design makes the most sense? Not sure? Well, let's find out.

Suppose you were responsible for designing a parking lot for a new office development. How would you set up the parking lot to maximize the number of parking spaces? In other words, how can you use the available area to get the most cars in? Follow the steps below to design your parking lot.

Considerations

Before you begin your design, you will have to consider some of the following issues:

- **How long** is a parking space?
- **How wide** is a parking space?
- **How wide** are the lines?
- How much distance has to be **between the rows** of parking spots?

Research those questions, then write the answers in the space provided.

How long ________________

How wide ________________

How wide are the lines ________________

Distance between rows ________________

What other issues do you have to consider?

____

____

____

____

(continued)

Name ______________________________ Date ____________________

18. Angled vs. Straight *(continued)*

Develop a Plan

The parking lot you are to redesign measures 250 feet by 175 feet, with three planters measuring 15 feet by 15 feet. How would you design the lot to maximize the number of spaces while meeting the city or county ordinances?

250 feet

175 feet

Explain what type of parking design is the most efficient.

____

____

____

____

Share Your Bright Ideas

We want to hear from you!

Your name____Date____________________

School name____

School address____

City ______________________________State _____Zip__________Phone number (______)__________

Grade level(s) taught________Subject area(s) taught_____________________________________

Where did you purchase this publication?____

In what month do you purchase a majority of your supplements?_________________________________

What moneys were used to purchase this product?

___School supplemental budget ___Federal/state funding ___Personal

Please "grade" this Walch publication in the following areas:

Quality of service you received when purchasing .... A B C D
Ease of use .... A B C D
Quality of content .... A B C D
Page layout .... A B C D
Organization of material .... A B C D
Suitability for grade level .... A B C D
Instructional value .... A B C D

COMMENTS:____

____

What specific supplemental materials would help you meet your current—or future—instructional needs?

____

Have you used other Walch publications? If so, which ones?_______________________________

May we use your comments in upcoming communications? ___Yes ___No

Please **FAX** this completed form to **888-991-5755**, or mail it to

Customer Service, Walch Publishing, P. O. Box 658, Portland, ME 04104-0658

We will send you a **FREE GIFT** in appreciation of your feedback. **THANK YOU!**